Introduction

Welcome to *Edexcel Topic Tutor*. This series of resources support learning and revision in the most popular A-level units of the Edexcel course.

This book contains a series of topic tests using real past exam questions, which can be tackled in class, as homework or as part of your revision. Short answers at the back of the book allow you to quickly spot problem areas. Where you need extra help, you can use the accompanying CD-ROM to view animated model answers with commentary from examiners.

The book also includes two unit exam papers for additional practice. Answers to these and other Edexcel examination papers are available at: www.examzone.com

Getting started

Insert the disk into your CD-ROM drive. Edexcel Topic Tutor should auto-run on your PC. If it does not auto run, click Start, My Computer, and right-click on the CD-ROM icon and choose 'Explore'. Double click on the Tutor.exe application and follow the onscreen instructions.

If you are a Mac user, you will need to double-click the CD icon on the desktop and double-click on the Topic Tutor application.

Animated model solutions are provided for all exercises. To see and hear a worked solution to a question you will need to:

1. Click on the exercise name from the menu of exercises (a numbered list of answers will appear)

2. Click the relevant answer number/part

This will open the Live Learning Player. You can click through the many steps of a solution or replay it using the control buttons at the bottom of the Live Learning Player.

You can view brief answers for any question by clicking on the 'Answers' link on the main screen.

You can view select past exam papers by clicking on the links at the bottom of the screen. You will need Adobe® Acrobat® Reader® installed on your machine or network in order to view both the answer PDFs and the sample paper PDFs. If you do not have Adobe® Acrobat® Reader®, follow the installation instructions on the Help screen.

Taking the Topic Tests

The number of marks available for a question is shown beside each question or question part.

You can check the answers to each question in the answer section at the end of each unit.

Using the exam papers

On the front page of each examination paper there are details about the examination you are taking, such as the amount of time in which you must answer the questions.

All questions have space in which to write your final answer and some have space in which to show your working.

Hardware Requirements

- Operating system: Windows 95(OS R2), 98, ME, 2000, NT or XP
- Pentium 400 (IBM Compatible PC) or equivalent PC
- 128 MB RAM or higher
- 16 bit graphics card
- CD-ROM drive (minimum 16 speed recommended)
- SVGA colour monitor and 1024/768 resolution
- Sound card
- At least 100 MB free hard disk space

Mac System requirements

- Operating system: X 10.1.5 or higher
- 500MHz G4 processor
- 256MB of RAM or higher
- 450MB of free hard disk space
- 16 speed CD ROM drive
- 16 bit colour monitor set at 1024/768 resolution

Technical support and advice is available from the Ask Edexcel service at www.edexcel.org.uk/ask.

Core Mathematics C3

Contents

Core Mathematics C3 - Formulae

Logarithms and exponentials

$$e^{x \ln a} = a^x$$

Trigonometric identities

$$\sin(A \pm B) = \sin A \cos B \pm \cos A \sin B$$

$$\cos(A \pm B) = \cos A \cos B \mp \sin A \sin B$$

$$\tan(A \pm B) = \frac{\tan A \pm \tan B}{1 \mp \tan A \tan B} \qquad (A \pm B \neq (k + \tfrac{1}{2})\pi)$$

$$\sin A + \sin B = 2 \sin \frac{A+B}{2} \cos \frac{A-B}{2}$$

$$\sin A - \sin B = 2 \cos \frac{A+B}{2} \sin \frac{A-B}{2}$$

$$\cos A + \cos B = 2 \cos \frac{A+B}{2} \cos \frac{A-B}{2}$$

$$\cos A - \cos B = -2 \sin \frac{A+B}{2} \sin \frac{A-B}{2}$$

Differentiation

$f(x)$	$f'(x)$
$\tan kx$	$k \sec^2 kx$
$\sec x$	$\sec x \tan x$
$\cot x$	$-\operatorname{cosec}^2 x$
$\operatorname{cosec} x$	$-\operatorname{cosec} x \cot x$
$\dfrac{f(x)}{g(x)}$	$\dfrac{f'(x)g(x) - f(x)g'(x)}{(g(x))^2}$

A - Algebra and functions

1. Express $\dfrac{2x^2 + 3x}{(2x + 3)(x - 2)} - \dfrac{6}{x^2 - x - 2}$ as a single fraction in its simplest form.

(7)

2. The function f is defined by $\text{f}: x \mapsto \dfrac{5x + 1}{x^2 + x - 2} - \dfrac{3}{x + 2}, \quad x > 1.$

(a) Show that $\text{f}(x) = \dfrac{2}{x - 1}, \ x > 1.$

(4)

(b) Find $\text{f}^{-1}(x)$.

(3)

The function g is defined by $\text{g}: x \mapsto x^2 + 5, \quad x \in \mathbb{R}.$

(c) Solve $\text{fg}(x) = \frac{1}{4}$.

(3)

3. The functions f and g are defined by

$$\text{f}: x \mapsto 2x + \ln 2, \ x \in \mathbb{R}, \ \text{g}: x \mapsto e^{2x}, \ x \in \mathbb{R}.$$

(a) Prove that the composite function gf is $\text{gf}: x \mapsto 4e^{4x}, \ x \in \mathbb{R}.$

(4)

(b) Sketch the curve with equation $y = \text{gf}(x)$, and show the coordinates of the point where the curve cuts the y-axis.

(1)

(c) Write down the range of gf.

(1)

(d) Find the value of x for which $\dfrac{\text{d}}{\text{d}x}[\text{gf}(x)] = 3$, giving your answer to 3 significant figures.

(4)

Core Mathematics C3

4. The function f is given by f: $x \mapsto \ln(3x - 6)$, $x \in \mathbb{R}$, $x > 2$.

(a) Find $f^{-1}(x)$.

(3)

(b) Write down the domain of f^{-1} and the range of f^{-1}.

(2)

(c) Find, to 3 significant figures, the value of x for which $f(x) = 3$.

(2)

5.

Figure 1

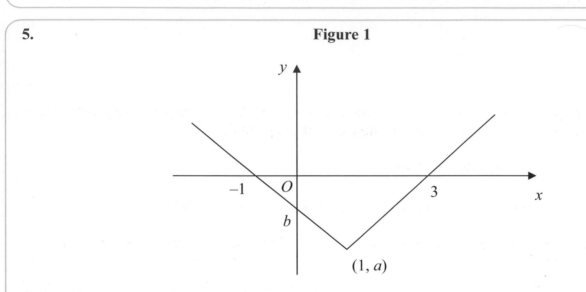

$(1, a)$

Figure 1 shows part of the graph of $y = f(x)$, $x \in \mathbb{R}$. The graph consists of two line segments that meet at the point $(1, a)$, $a < 0$. One line meets the x-axis at $(3, 0)$. The other line meets the x-axis at $(-1, 0)$ and the y-axis at $(0, b)$, $b < 0$.

In separate diagrams, sketch the graph with equation

(a) $y = f(x + 1)$,   **(2)**   (b) $y = f(|x|)$.   **(3)**

Indicate clearly on each sketch the coordinates of any points of intersection with the axes.

Given that $f(x) = |x - 1| - 2$, find

(c) the value of a and the value of b,

(2)

(d) the value of x for which $f(x) = 5x$.

(4)

6.   **Figure 1**

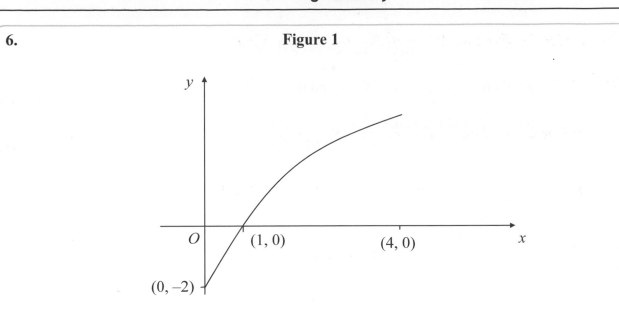

Figure 1 shows a sketch of the curve with equation $y = f(x)$, $0 \leq x \leq 4$. The curve passes through the point $(1, 0)$ on the x-axis and meets the y-axis at the point $(0, -2)$.

Sketch, on separate axes, the graph of

(a) $y = |f(x)|$,   **(2)**   (b) $y = f(2x)$,   **(2)**   (c) $y = f^{-1}(x)$,   **(3)**

in each case showing the coordinates of the points at which the graph meets the axes.

B - Trigonometry

1.   (a) Given that $\sin^2 \theta + \cos^2 \theta \equiv 1$, show that $1 + \tan^2 \theta \equiv \sec^2 \theta$.

(2)

   (b) Solve, for $0 \leq \theta < 360°$, the equation $2\tan^2 \theta + \sec \theta = 1$, giving your answers to 1 decimal place.

(6)

2.   (a) Given that $\sin x = \dfrac{3}{5}$, use an appropriate double angle formula to find the exact value of $\sec 2x$.

(4)

   (b) Prove that $\cot 2x + \operatorname{cosec} 2x \equiv \cot x,$ $\left(x \neq \dfrac{n\pi}{2}, n \in \mathbb{Z} \right).$

(4)

3.   (a) Given that $2\sin(\theta + 30)° = \cos(\theta + 60)°$, find the exact value of $\tan \theta°$.

(5)

   (b) (i)   Using the identity $\cos(A + B) \equiv \cos A \cos B - \sin A \sin B$, prove that
            $\cos 2A \equiv 1 - 2\sin^2 A$.

(2)

       (ii)   Hence solve, for $0 \leq x < 2\pi$, $\cos 2x = \sin x$,
            giving your answers in terms of π.

(5)

       (iii)   Show that $\sin 2y \tan y + \cos 2y \equiv 1$, for $0 \leq y < \frac{1}{2}\pi$.

(3)

4.
$$f(x) = 12 \cos x - 4 \sin x.$$

Given that $f(x) = R \cos (x + \alpha)$, where $R \geq 0$ and $0 \leq \alpha \leq 90°$,

(*a*) find the value of R and the value of α.

(4)

(*b*) Hence solve the equation $12 \cos x - 4 \sin x = 7$ for $0 \leq x < 360°$, giving your answers to one decimal place.

(5)

(*c*) (i) Write down the minimum value of $12 \cos x - 4 \sin x$.

(1)

(ii) Find, to 2 decimal places, the smallest positive value of x for which this minimum value occurs.

(2)

5. (*a*) Show that $2 \sin 2\theta - 3 \cos 2\theta - 3 \sin \theta + 3 \equiv \sin \theta (4 \cos \theta + 6 \sin \theta - 3)$.

(4)

(*b*) Express $4 \cos \theta + 6 \sin \theta$ in the form $R \sin (\theta + \alpha)$, where $R > 0$ and $0 < \alpha < \frac{1}{2}\pi$.

(4)

(*c*) Hence, for $0 \leq \theta < \pi$, solve $2 \sin 2\theta = 3(\cos 2\theta + \sin \theta - 1)$, giving your answers in radians to 3 significant figures, where appropriate.

(5)

6. (*a*) Show that

(i) $\dfrac{\cos 2x}{\cos x + \sin x} \equiv \cos x - \sin x, \ x \neq (n - \frac{1}{4})\pi, \ n \in \mathbb{Z},$

(2)

(ii) $\frac{1}{2} (\cos 2x - \sin 2x) \equiv \cos^2 x - \cos x \sin x - \frac{1}{2}.$

(3)

(*b*) Hence, or otherwise, show that the equation

$$\cos \theta \left(\frac{\cos 2\theta}{\cos \theta + \sin \theta} \right) = \frac{1}{2}$$

can be written as $\sin 2\theta = \cos 2\theta$.

(3)

(*c*) Solve, for $0 \leq \theta < 2\pi$, $\sin 2\theta = \cos 2\theta$, giving your answers in terms of π.

(4)

C - Exponentials and logarithms

1. A particular species of orchid is being studied. The population p at time t years after the study started is assumed to be $p = \dfrac{2800ae^{0.2t}}{1 + ae^{0.2t}}$, where a is a constant.

Given that there were 300 orchids when the study started,

(a) show that $a = 0.12$,

(3)

(b) use the equation with $a = 0.12$ to predict the number of years before the population of orchids reaches 1850.

(4)

(c) Show that $p = \dfrac{336}{0.12 + e^{-0.2t}}$.

(1)

(d) Hence show that the population cannot exceed 2800.

(2)

2. Find the exact solutions of   (i) $e^{2x + 3} = 6$,   (3)   (ii) $\ln(3x + 2) = 4$.   (3)

3. As a substance cools its temperature, $T\,°C$, is related to the time (t minutes) for which it has been cooling. The relationship is given by the equation $T = 20 + 60e^{-0.1t}$, $t \geq 0$.

(a) Find the value of T when the substance started to cool.

(1)

(b) Explain why the temperature of the substance is always above 20°C.

(1)

(c) Sketch the graph of T against t.

(2)

(d) Find the value, to 2 significant figures, of t at the instant $T = 60$.

(4)

D - Differentiation

1. Use the derivatives of cosec x and cot x to prove that $\dfrac{d}{dx}\left[\ln(\operatorname{cosec} x + \cot x)\right] = -\operatorname{cosec} x.$

(3)

Core Mathematics C3

2. (a) Differentiate with respect to x

    (i) $3\sin^2 x + \sec 2x$, **(3)**   (ii) $\{x + \ln(2x)\}^3$. **(3)**

Given that $y = \dfrac{5x^2 - 10x + 9}{(x-1)^2}$, $x \neq 1$,

(b) show that $\dfrac{dy}{dx} = -\dfrac{8}{(x-1)^3}$.

(6)

3. The point P lies on the curve with equation $y = \ln\left(\dfrac{1}{3}x\right)$. The x-coordinate of P is 3.

Find an equation of the normal to the curve at the point P in the form $y = ax + b$, where a and b are constants.

(5)

4. (a) Differentiate with respect to x (i) $x^2 e^{3x+2}$, **(4)**   (ii) $\dfrac{\cos(2x^3)}{3x}$. **(4)**

(b) Given that $x = 4\sin(2y + 6)$, find $\dfrac{dy}{dx}$ in terms of x.

(5)

5. $$f(x) = (x^2 + 1)\ln x, \quad x > 0.$$

Use differentiation to find the value of $f'(x)$ at $x = e$, leaving your answer in terms of e.

(4)

E - Numerical methods

1. $$f(x) = 2x^3 - x - 4.$$

(a) Show that the equation $f(x) = 0$ can be written as $x = \sqrt{\left(\dfrac{2}{x} + \dfrac{1}{2}\right)}$.

(3)

The equation $2x^3 - x - 4 = 0$ has a root between 1.35 and 1.4.

(b) Use the iteration formula $\quad x_{n+1} = \sqrt{\left(\dfrac{2}{x_n} + \dfrac{1}{2}\right)}$,

with $x_0 = 1.35$, to find, to 2 decimal places, the value of x_1, x_2 and x_3.

(3)

The only real root of $f(x) = 0$ is α.

(c) By choosing a suitable interval, prove that $\alpha = 1.392$, to 3 decimal places.

(3)

2.

Figure 4

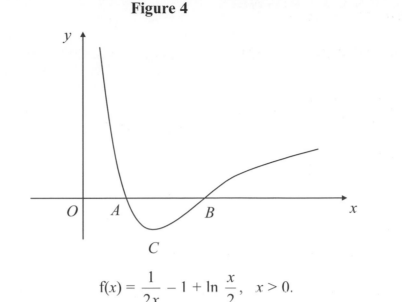

$$f(x) = \frac{1}{2x} - 1 + \ln \frac{x}{2}, \quad x > 0.$$

Figure 4 shows part of the curve with equation $y = f(x)$. The curve crosses the x-axis at the points A and B, and has a minimum at the point C.

(a) Show that the x-coordinate of C is $\frac{1}{2}$.

(5)

(b) Find the y-coordinate of C in the form $k \ln 2$, where k is a constant.

(2)

(c) Verify that the x-coordinate of B lies between 4.905 and 4.915.

(2)

(d) Show that the equation $\dfrac{1}{2x} - 1 + \ln \dfrac{x}{2} = 0$ can be rearranged into the form $x = 2e^{\left(1 - \frac{1}{2x}\right)}$.

(2)

The x-coordinate of B is to be found using the iterative formula $x_{n+1} = 2e^{\left(1 - \frac{1}{2x_n}\right)}$, with $x_0 = 5$.

(e) Calculate, to 4 decimal places, the values of x_1, x_2 and x_3.

(2)

3.
$$f(x) = 3e^x - \tfrac{1}{2} \ln x - 2, \quad x > 0.$$

(a) Differentiate to find $f'(x)$.

(3)

The curve with equation $y = f(x)$ has a turning point at P. The x-coordinate of P is α.

(b) Show that $\alpha = \tfrac{1}{6} e^{-\alpha}$.

(2)

The iterative formula $x_{n+1} = \tfrac{1}{6} e^{-x_n}$, $x_0 = 1$, is used to find an approximate value for α.

(c) Calculate the values of x_1, x_2, x_3 and x_4, giving your answers to 4 decimal places. (2)

(d) By considering the change of sign of $f'(x)$ in a suitable interval, prove that $\alpha = 0.1443$ correct to 4 decimal places.

(2)

Core Mathematics C3 - Answers

A - Algebra and functions

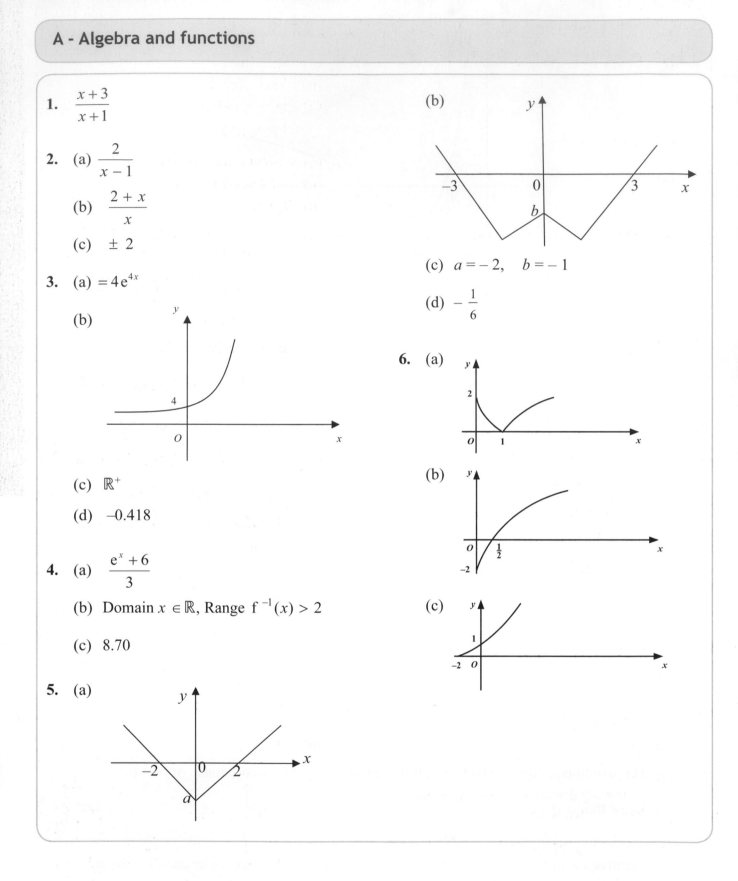

1. $\dfrac{x+3}{x+1}$

2. (a) $\dfrac{2}{x-1}$

   (b) $\dfrac{2+x}{x}$

   (c) ± 2

3. (a) $=4e^{4x}$

   (b)

   (c) $\mathbb{R}^{+}$

   (d) -0.418

4. (a) $\dfrac{e^{x}+6}{3}$

   (b) Domain $x \in \mathbb{R}$, Range $f^{-1}(x) > 2$

   (c) 8.70

5. (a)

(b)

   (c) $a = -2, \quad b = -1$

   (d) $-\dfrac{1}{6}$

6. (a)

   (b)

   (c)

B - Trigonometry

1. (a) Divide by $\cos^2\theta$ gives $\tan^2\theta + 1 =$

$$\frac{1}{\cos^2\theta}$$

(b) $0, 131.8, 228.2$

2. (a) $\dfrac{25}{7}$

(b) $\cot 2x + \operatorname{cosec} 2x \approx \dfrac{\cos 2x}{\sin 2x} + \dfrac{1}{\sin 2x}$

$\approx \dfrac{2\cos^2 x - 1 + 1}{\sin 2x} \approx \dfrac{2\cos^2 x}{2\sin x \cos x} \approx \cot x$

3. (a) $-\dfrac{1}{3\sqrt{3}}$

(b) (i) Set $A = B$ to give

$$\cos 2A \approx \cos^2 A - \sin^2 A$$
$$\approx (1 - \sin^2 A) - \sin^2 A$$
$$\approx 1 - 2\sin^2 A$$

(ii) $\dfrac{\pi}{6}, \dfrac{5\pi}{6}, \dfrac{3\pi}{2}$

(iii) $\sin 2y \tan y + \cos 2y$

$\approx 2\sin y \cos y \dfrac{\sin y}{\cos y} + (1 - 2\sin^2 y)$

$\approx 2\sin^2 y + (1 - 2\sin^2 y) \approx 1$

4. (a) $12.6, 18.4°$

(b) $38.0°, 285.2°$

(c) (i) $-\sqrt{160}$ (-12.6)

(ii) $161.57°$

5. (a) $\sin\theta\,(4\cos\theta + 6\sin\theta - 3)$

(b) $\sqrt{52}\sin(\theta + 0.588)$

(c) $0, 2.12$

6. (a) (i) $\dfrac{\cos 2x}{\cos x + \sin x} \approx \dfrac{\cos^2 x - \sin^2 x}{\cos x + \sin x}$

$\approx \dfrac{(\cos x - \sin x)(\cos x + \sin x)}{\cos x + \sin x}$

$\approx \cos x - \sin x$

(ii) $\dfrac{1}{2}(\cos 2x - \sin 2x)$

$\approx \dfrac{1}{2}(2\cos^2 x - 1 - 2\sin x \cos x)$

$\approx \cos^2 x - \cos x \sin x - \dfrac{1}{2}$

(b) $\cos\theta\,(\cos\theta - \sin\theta) = \dfrac{1}{2}$ Using (a) (i)

$\cos^2\theta - \cos\theta\sin\theta - \dfrac{1}{2} = 0$

$\dfrac{1}{2}(\cos 2\theta - \sin 2\theta) = 0$ Using (a) (ii)

$\cos 2\theta = \sin 2\theta$

(c) $\dfrac{\pi}{8}, \dfrac{5\pi}{8}, \dfrac{9\pi}{8}, \dfrac{13\pi}{8}$

C - Exponentials and logarithms

1. (a) $\dfrac{2800a}{1+a} = 300$

(b) 14

(c) Correct derivation:
(Showing division of num. and den.
by $e^{0.2t}$; using (a)

(d) $e^{-0.2t} \geqslant 0$ so limit $\dfrac{336}{0.12}$

2. (i) $\dfrac{1}{2}(\ln 6 - 3)$

(ii) $\dfrac{1}{3}(e^4 - 2)$

3. (a) 80

(b) $e^{-0.1\,t} \geq 0$

(c)

(d) 4.1

D - Differentiation

1. $\dfrac{1}{\operatorname{cosec} x + \cot x}(-\cot x \operatorname{cosec} x - \operatorname{cosec}^2 x)$

$= -\operatorname{cosec} x$

2. (a) (i) $6\sin x \cos x + 2\sec 2x \tan 2x$

    (ii) $3(x + \ln 2x)^2(1 + \dfrac{1}{x})$

  (b) $-\dfrac{(x-1)^3}{8}$

3. $y = -3x + 9$

4. (a) (i) $3x^2 e^{3x+2} + 2x e^{3x+2}$

    (ii) $\dfrac{-18x^3 \sin(2x^3) - 3\cos(2x^3)}{9x^2}$

  (b) $\dfrac{1}{8\cos\left(\arcsin\left(\dfrac{x}{4}\right)\right)}$

5. $3e + \dfrac{1}{e}$

E - Numerical Methods

1. (a) $0 = x^2 - \dfrac{1}{2} - \dfrac{2}{x}$

    $x = \sqrt{\left(\dfrac{2}{x} + \dfrac{1}{2}\right)}$

  (b) 1.41, 1.39, 1.39

  (c) $f(1.3915) \approx -3\times10^{-3}$,

    $f(1.3925) \approx 7\times10^{-3}$

2. (a) $f'(x) = \dfrac{-1}{2x^2} + \dfrac{1}{x} = 0$

    $x = 0.5$

  (b) $-2\ln 2$

  (c) $f(4.905) = < 0 \ (-0.000955)$,
    $f(4.915) = > 0 \ (+0.000874)$

  (d) $\ln\left(\dfrac{x}{2}\right) = 1 - \dfrac{1}{2x}$

    $\dfrac{x}{2} = e^{\left(1 - \frac{1}{2x}\right)}$

    $x = 2e^{\left(1 - \frac{1}{2x}\right)}$

  (e) 4.9192, 4.9111, 4.9103

3. (a) $3e^x - \dfrac{1}{2x}$

  (b) $\dfrac{1}{6}e^{-\alpha}$

  (c) $x_1 = 0.0613...,\ x_2 = 0.1568..,$
    $x_3 = 0.1425...,\ x_4 = 0.1445....$

  (d) $f'(0.14435) = +0.002(1)$
    $f'(0.14425) = -0.007$

Core Mathematics C3 - Practice paper

Answers to these practice papers have not been included in this book. Your teacher should have access to the mark schemes through Edexcel Online. Alternatively, visit www.examzone.co.uk to download model answers.

Paper Reference(s)

6665/01

Edexcel GCE

Core Mathematics C3

Advanced Level

Monday 12 June 2006 – Afternoon

Time: 1 hour 30 minutes

Materials required for examination	Items included with question papers
Mathematical Formulae (Green)	Nil

Candidates may use any calculator EXCEPT those with the facility for symbolic algebra, differentiation and/or integration. Thus candidates may NOT use calculators such as the Texas Instruments TI 89, TI 92, Casio CFX 9970G, Hewlett Packard HP 48G.

Instructions to Candidates

In the boxes above, write your centre number, candidate number, your surname, initial(s) and signature.
Check that you have the correct question paper.
When a calculator is used, the answer should be given to an appropriate degree of accuracy.
You must write your answer for each question in the space following the question.
If you need more space to complete your answer to any question, use additional answer sheets.

Information for Candidates

The marks for individual questions and the parts of questions are shown in round brackets: e.g. **(2)**.
Full marks may be obtained for answers to ALL questions.
There are 8 questions in this question paper. The total mark for this paper is 75.

Advice to Candidates

You must ensure that your answers to parts of questions are clearly labelled.
You must show sufficient working to make your methods clear to the Examiner. Answers without working may gain no credit.

1. (a) Simplify $\dfrac{3x^2 - x - 2}{x^2 - 1}$.

(3)

(b) Hence, or otherwise, express $\dfrac{3x^2 - x - 2}{x^2 - 1} - \dfrac{1}{x(x+1)}$ as a single fraction in its simplest form.

(3)

1. (a) Simplify $\dfrac{3x^2 - x - 2}{x^2 - 1}$

2. Differentiate, with respect to x,

(a) $e^{3x} + \ln 2x$,

(3)

(b) $(5 + x^2)^{\frac{3}{2}}$.

(3)

3.   **Figure 1**

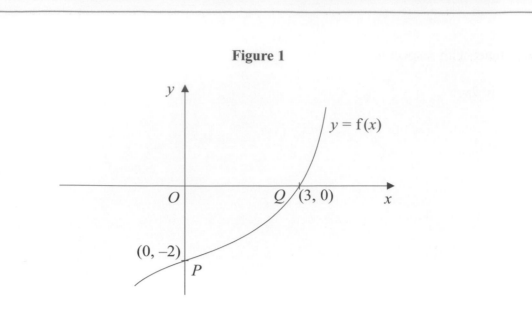

Figure 1 shows part of the curve with equation $y = f(x)$, $x \in \mathbb{R}$, where f is an increasing function of x. The curve passes through the points $P(0, -2)$ and $Q(3, 0)$ as shown.

In separate diagrams, sketch the curve with equation

(a) $y = |f(x)|$,

(3)

(b) $y = f^{-1}(x)$,

(3)

(c) $y = \frac{1}{2} f(3x)$.

(3)

Indicate clearly on each sketch the coordinates of the points at which the curve crosses or meets the axes.

Question 3 continued

Question 3 continued

4.　A heated metal ball is dropped into a liquid. As the ball cools, its temperature, $T\,°C$, t minutes after it enters the liquid, is given by

$$T = 400\,e^{-0.05t} + 25, \quad t \geqslant 0.$$

(a)　Find the temperature of the ball as it enters the liquid.

(1)

(b)　Find the value of t for which $T = 300$, giving your answer to 3 significant figures.

(4)

(c)　Find the rate at which the temperature of the ball is decreasing at the instant when $t = 50$. Give your answer in $°C$ per minute to 3 significant figures.

(3)

(d)　From the equation for temperature T in terms of t, given above, explain why the temperature of the ball can never fall to $20\,°C$.

(1)

Leave blank

Question 4 continued

Question 4 continued

5.

Figure 2

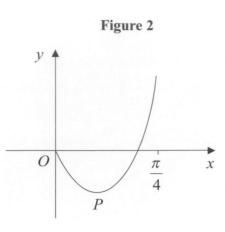

Figure 2 shows part of the curve with equation

$$y = (2x-1)\tan 2x, \quad 0 \leqslant x < \frac{\pi}{4}.$$

The curve has a minimum at the point P. The x-coordinate of P is k.

(a) Show that k satisfies the equation

$$4k + \sin 4k - 2 = 0.$$

(6)

The iterative formula

$$x_{n+1} = \frac{1}{4}(2 - \sin 4x_n), \quad x_0 = 0.3,$$

is used to find an approximate value for k.

(b) Calculate the values of x_1, x_2, x_3 and x_4, giving your answers to 4 decimal places.

(3)

(c) Show that $k = 0.277$, correct to 3 significant figures.

(2)

Leave blank

Question 5 continued

6. (a) Using $\sin^2\theta + \cos^2\theta \equiv 1$, show that $\operatorname{cosec}^2\theta - \cot^2\theta \equiv 1$.

(2)

(b) Hence, or otherwise, prove that

$$\operatorname{cosec}^4\theta - \cot^4\theta \equiv \operatorname{cosec}^2\theta + \cot^2\theta.$$

(2)

(c) Solve, for $90° < \theta < 180°$,

$$\operatorname{cosec}^4\theta - \cot^4\theta = 2 - \cot\theta.$$

(6)

6. (a) Using $\sin^2\theta + \cos^2\theta \equiv 1$, show that $\operatorname{cosec}^2\theta - \cot^2\theta \equiv 1$.

Leave
blank

Question 6 continued

7.   For the constant k, where $k > 1$, the functions f and g are defined by

$$\text{f}: x \longmapsto \ln(x + k), \quad x > -k,$$
$$\text{g}: x \longmapsto |2x - k|, \quad x \in \mathbb{R}.$$

(a)   On separate axes, sketch the graph of f and the graph of g.

On each sketch state, in terms of k, the coordinates of points where the graph meets the coordinate axes.

(5)

(b)   Write down the range of f.

(1)

(c)   Find $\text{fg}\left(\dfrac{k}{4}\right)$ in terms of k, giving your answer in its simplest form.

(2)

The curve C has equation $y = \text{f}(x)$. The tangent to C at the point with x-coordinate 3 is parallel to the line with equation $9y = 2x + 1$.

(d)   Find the value of k.

(4)

Leave
blank

Question 7 continued

Question 7 continued

Question 7 continued

8. (a) Given that $\cos A = \dfrac{3}{4}$, where $270° < A < 360°$, find the exact value of $\sin 2A$.

(5)

(b) (i) Show that $\cos\left(2x + \dfrac{\pi}{3}\right) + \cos\left(2x - \dfrac{\pi}{3}\right) \equiv \cos 2x$.

(3)

Given that

$$y = 3\sin^2 x + \cos\left(2x + \frac{\pi}{3}\right) + \cos\left(2x - \frac{\pi}{3}\right),$$

(ii) show that $\dfrac{dy}{dx} = \sin 2x$.

(4)

Question 8 continued

Leave
blank

Question 8 continued

TOTAL FOR PAPER: 75 MARKS

END

Core Mathematics C4

Contents

Core Mathematics C4 - Formulae

Integration (+ constant)

$f(x)$	$\int f(x) \; dx$
$\sec^2 kx$	$\dfrac{1}{k} \tan kx$
$\tan x$	$\ln\left|\sec x\right|$
$\cot x$	$\ln\left|\sin x\right|$
$\operatorname{cosec} x$	$-\ln\left|\operatorname{cosec} x + \cot x\right| = \ln\left|\tan(\tfrac{1}{2} x)\right|$
$\sec x$	$\ln\left|\sec x + \tan x\right| = \ln\left|\tan(\tfrac{1}{2} x + \tfrac{1}{4}\pi)\right|$

$$\int u \frac{dv}{dx} \, dx = uv - \int v \frac{du}{dx} \, dx$$

A - Algebra and functions

1. Given that $\dfrac{3+5x}{(1+3x)(1-x)} \equiv \dfrac{A}{1+3x} + \dfrac{B}{1-x}$, find the values of the constants A and B.

(3)

2. $\mathrm{f}(x) = \dfrac{3x^2 + 16}{(1-3x)(2+x)^2} = \dfrac{A}{(1-3x)} + \dfrac{B}{(2+x)} + \dfrac{C}{(2+x)^2}, \quad |x| < \tfrac{1}{3}.$

Find the values of A and C and show that $B = 0$.

(4)

B - Coordinate geometry

1.

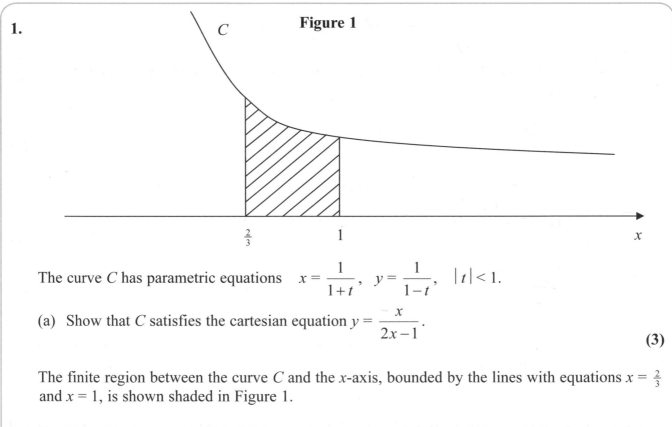

Figure 1

The curve C has parametric equations $\quad x = \dfrac{1}{1+t}, \quad y = \dfrac{1}{1-t}, \quad |t| < 1.$

(a) Show that C satisfies the cartesian equation $y = \dfrac{x}{2x-1}.$

(3)

The finite region between the curve C and the x-axis, bounded by the lines with equations $x = \tfrac{2}{3}$ and $x = 1$, is shown shaded in Figure 1.

(b) Calculate the exact value of the area of this region, giving your answer in the form $a + b \ln c$, where a, b and c are constants.

(6)

2. A curve has parametric equations $x = 2 \cot t, \quad y = 2 \sin^2 t, \quad 0 < t \le \dfrac{\pi}{2}$.

   Find a cartesian equation of the curve in the form $y = f(x)$. State the domain on which the curve is defined.

   (4)

3.

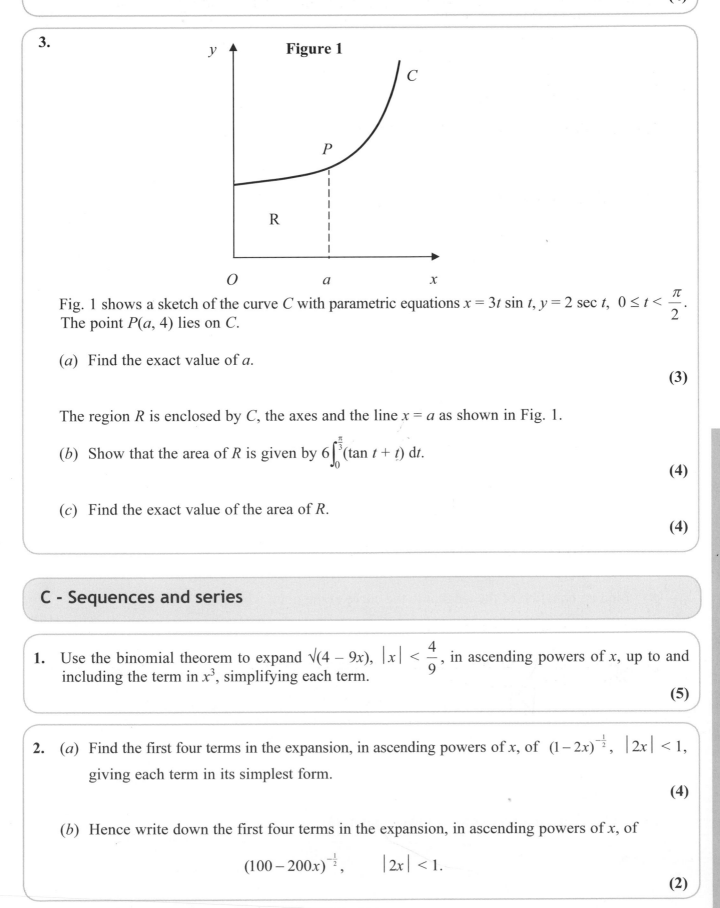

Fig. 1 shows a sketch of the curve C with parametric equations $x = 3t \sin t, y = 2 \sec t, 0 \le t < \dfrac{\pi}{2}$. The point $P(a, 4)$ lies on C.

(a) Find the exact value of a.

   (3)

The region R is enclosed by C, the axes and the line $x = a$ as shown in Fig. 1.

(b) Show that the area of R is given by $6\displaystyle\int_0^{\frac{\pi}{3}}(\tan t + t)\, dt$.

   (4)

(c) Find the exact value of the area of R.

   (4)

C - Sequences and series

1. Use the binomial theorem to expand $\sqrt{(4 - 9x)}, |x| < \dfrac{4}{9}$, in ascending powers of x, up to and including the term in x^3, simplifying each term.

   (5)

2. (a) Find the first four terms in the expansion, in ascending powers of x, of $(1 - 2x)^{-\frac{1}{2}}, |2x| < 1$, giving each term in its simplest form.

   (4)

   (b) Hence write down the first four terms in the expansion, in ascending powers of x, of

   $$(100 - 200x)^{-\frac{1}{2}}, \qquad |2x| < 1.$$

   (2)

3. In the binomial expansion, in ascending powers of x, of $(1 + ax)^n$, where a and n are constants, the coefficient of x is 15. The coefficient of x^2 and of x^3 are equal.

(a) Find the value of a and the value of n.

(6)

(b) Find the coefficient of x^3.

(1)

4.
$$f(x) = \frac{1}{\sqrt{(1-x)}} - \sqrt{(1+x)}, \quad -1 < x < 1.$$

(a) Find the series expansion of $f(x)$, in ascending powers of x, up to and including the term in x^3.

(6)

(b) Hence, or otherwise, prove that the function f has a minimum at the origin.

(4)

D - Differentiation

1. A curve C is described by the equation $3x^2 + 4y^2 - 2x + 6xy - 5 = 0$.

Find an equation of the tangent to C at the point $(1, -2)$, giving your answer in the form $ax + by + c = 0$, where a, b and c are integers.

(7)

2. A curve has parametric equations $x = 2 \cot t, \quad y = 2 \sin^2 t, \quad 0 < t \le \frac{\pi}{2}$.

(a) Find an expression for $\frac{dy}{dx}$ in terms of the parameter t.

(4)

(b) Find an equation of the tangent to the curve at the point where $t = \frac{\pi}{4}$.

(4)

3. The value £V of a car t years after the 1st January 2001 is given by the formula $V = 10\,000 \times (1.5)^{-t}$.

(a) Find the value of the car on 1st January 2005.

(2)

(b) Find the value of $\frac{dV}{dt}$ when $t = 4$.

(3)

(c) Explain what the answer to part (b) represents.

(1)

4.

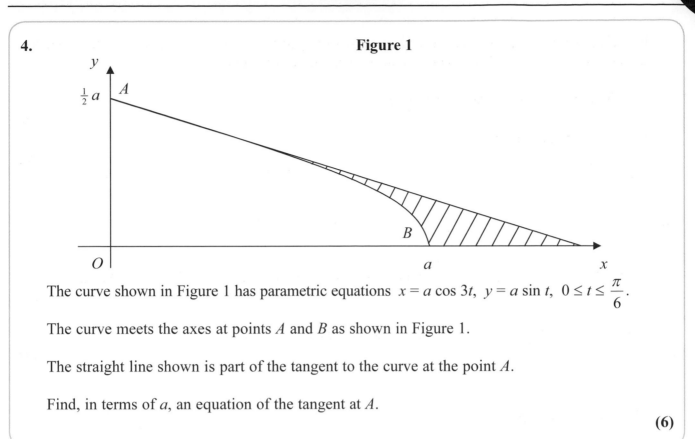

Figure 1

The curve shown in Figure 1 has parametric equations $x = a \cos 3t$, $y = a \sin t$, $0 \le t \le \dfrac{\pi}{6}$.

The curve meets the axes at points A and B as shown in Figure 1.

The straight line shown is part of the tangent to the curve at the point A.

Find, in terms of a, an equation of the tangent at A.

(6)

E - Integration

1. Using the substitution $u^2 = 2x - 1$, or otherwise, find the exact value of $\displaystyle\int_1^5 \dfrac{3x}{\sqrt{(2x-1)}} \, dx$.

(8)

2.
$$f(x) = \dfrac{9 + 4x^2}{9 - 4x^2}, \quad x \ne \pm\tfrac{3}{2}.$$

(a) Find the values of the constants A, B and C such that

$$f(x) = A + \dfrac{B}{3 + 2x} + \dfrac{C}{3 - 2x}, \quad x \ne \pm\tfrac{3}{2}.$$

(4)

(b) Hence find the exact value of $\displaystyle\int_{-1}^{1} \dfrac{9 + 4x^2}{9 - 4x^2} \, dx$.

(5)

3.
$$f(x) = (x^2 + 1) \ln x, \, x > 0.$$

Find the exact value of $\displaystyle\int_1^e f(x) \, dx$.

(5)

Core Mathematics C4

4. *(a)* Use integration by parts to find $\displaystyle\int x\cos 2x\ dx$.

(4)

(b) Hence, or otherwise, find $\displaystyle\int x\cos^2 x\ dx$.

(3)

5.

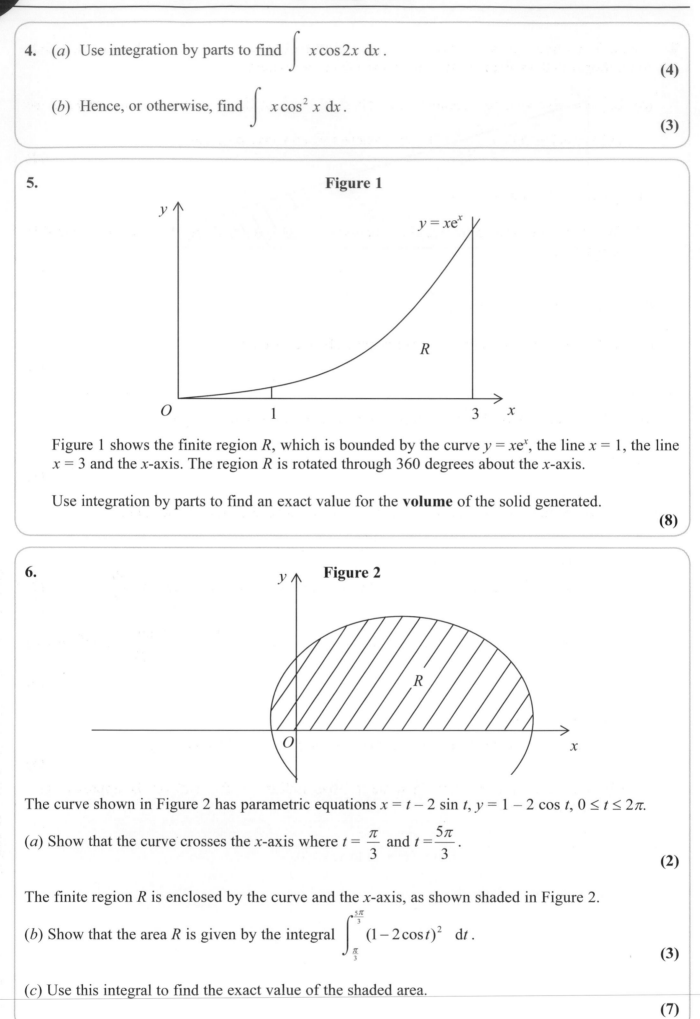

Figure 1

Figure 1 shows the finite region R, which is bounded by the curve $y = xe^x$, the line $x = 1$, the line $x = 3$ and the x-axis. The region R is rotated through 360 degrees about the x-axis.

Use integration by parts to find an exact value for the **volume** of the solid generated.

(8)

6.

Figure 2

The curve shown in Figure 2 has parametric equations $x = t - 2\sin t$, $y = 1 - 2\cos t$, $0 \le t \le 2\pi$.

(a) Show that the curve crosses the x-axis where $t = \dfrac{\pi}{3}$ and $t = \dfrac{5\pi}{3}$.

(2)

The finite region R is enclosed by the curve and the x-axis, as shown shaded in Figure 2.

(b) Show that the area R is given by the integral $\displaystyle\int_{\frac{\pi}{3}}^{\frac{5\pi}{3}} (1 - 2\cos t)^2\ dt$.

(3)

(c) Use this integral to find the exact value of the shaded area.

(7)

7. Liquid is pouring into a container at a constant rate of 20 cm³ s⁻¹ and is leaking out at a rate proportional to the volume of the liquid already in the container.

(a) Explain why, at time t seconds, the volume, V cm³, of liquid in the container satisfies the differential equation $\dfrac{dV}{dt} = 20 - kV$, where k is a positive constant.

(2)

The container is initially empty.

(b) By solving the differential equation, show that $V = A + Be^{-kt}$, giving the values of A and B in terms of k.

(6)

Given also that $\dfrac{dV}{dt} = 10$ when $t = 5$,

(c) find the volume of liquid in the container at 10 s after the start.

(5)

8. The volume of a spherical balloon of radius r cm is V cm³, where $V = \frac{4}{3}\pi r^3$.

(a) Find $\dfrac{dV}{dr}$.

(1)

The volume of the balloon increases with time t seconds according to the formula

$$\frac{dV}{dt} = \frac{1000}{(2t+1)^2}, \quad t \geq 0.$$

(b) Using the chain rule, or otherwise, find an expression in terms of r and t for $\dfrac{dr}{dt}$.

(2)

(c) Given that $V = 0$ when $t = 0$, solve the differential equation $\dfrac{dV}{dt} = \dfrac{1000}{(2t+1)^2}$, to obtain V in terms of t.

(4)

(d) Hence, at time $t = 5$,

(i) find the radius of the balloon, giving your answer to 3 significant figures,

(3)

(ii) show that the rate of increase of the radius of the balloon is approximately 2.90×10^{-2} cms⁻¹.

(2)

9. (a) Given that $y = \sec x$, complete the table with the values of y corresponding to

$$x = \frac{\pi}{16}, \frac{\pi}{8} \text{ and } \frac{\pi}{4}.$$

x	0	$\frac{\pi}{16}$	$\frac{\pi}{8}$	$\frac{3\pi}{16}$	$\frac{\pi}{4}$
y	1			1.20269	

(2)

(b) Use the trapezium rule, with all the values for y in the completed table, to obtain an estimate

for $\int_0^{\frac{\pi}{4}} \sec x \; dx$. Show all the steps of your working and give your answer to 4 decimal

places.

(3)

The exact value of $\int_0^{\frac{\pi}{4}} \sec x \; dx$ is $\ln(1 + \sqrt{2})$.

(c) Calculate the % error in using the estimate you obtained in part (b).

(2)

F - Vectors

1. The line l_1 has vector equation $\mathbf{r} = 8\mathbf{i} + 12\mathbf{j} + 14\mathbf{k} + \lambda(\mathbf{i} + \mathbf{j} - \mathbf{k})$, where λ is a parameter.

The point A has coordinates $(4, 8, a)$, where a is a constant. The point B has coordinates $(b, 13, 13)$, where b is a constant. Points A and B lie on the line l_1.

(a) Find the values of a and b.

(3)

Given that the point O is the origin, and that the point P lies on l_1 such that OP is perpendicular to l_1,

(b) find the coordinates of P.

(5)

(c) Hence find the distance OP, giving your answer as a simplified surd.

(2)

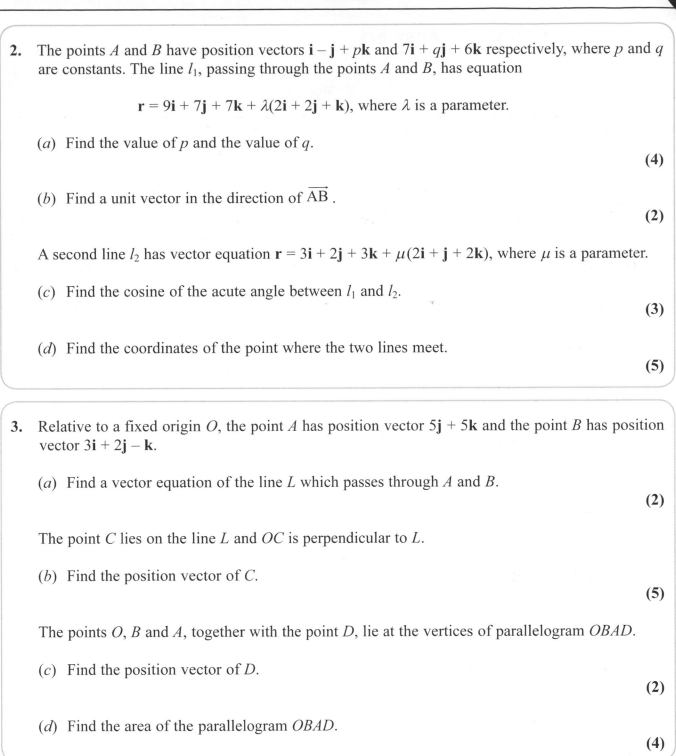

2. The points A and B have position vectors $\mathbf{i} - \mathbf{j} + p\mathbf{k}$ and $7\mathbf{i} + q\mathbf{j} + 6\mathbf{k}$ respectively, where p and q are constants. The line l_1, passing through the points A and B, has equation

$$\mathbf{r} = 9\mathbf{i} + 7\mathbf{j} + 7\mathbf{k} + \lambda(2\mathbf{i} + 2\mathbf{j} + \mathbf{k}), \text{ where } \lambda \text{ is a parameter.}$$

(a) Find the value of p and the value of q.

(4)

(b) Find a unit vector in the direction of $\overrightarrow{AB}$.

(2)

A second line l_2 has vector equation $\mathbf{r} = 3\mathbf{i} + 2\mathbf{j} + 3\mathbf{k} + \mu(2\mathbf{i} + \mathbf{j} + 2\mathbf{k})$, where μ is a parameter.

(c) Find the cosine of the acute angle between l_1 and l_2.

(3)

(d) Find the coordinates of the point where the two lines meet.

(5)

3. Relative to a fixed origin O, the point A has position vector $5\mathbf{j} + 5\mathbf{k}$ and the point B has position vector $3\mathbf{i} + 2\mathbf{j} - \mathbf{k}$.

(a) Find a vector equation of the line L which passes through A and B.

(2)

The point C lies on the line L and OC is perpendicular to L.

(b) Find the position vector of C.

(5)

The points O, B and A, together with the point D, lie at the vertices of parallelogram $OBAD$.

(c) Find the position vector of D.

(2)

(d) Find the area of the parallelogram $OBAD$.

(4)

Core Mathematics C4 - Answers

A - Algebra and functions

1. $A = 1, B = 2$

2. $A = 3, C = 4$

B - Coordinate geometry

1. (a) $\dfrac{x}{2x-1} = \dfrac{\frac{1}{(1+t)}}{\frac{2}{1+t}-1} = \dfrac{1}{2-(1+t)} = \dfrac{1}{1-t} = y$

   (b) $\dfrac{1}{6} + \dfrac{1}{4}\ln 3$

2. $y = \dfrac{8}{4+x^2}, \ x \geq 0$

3. (a) $\dfrac{\pi\sqrt{3}}{2}$

   (b) $\text{Area} = \displaystyle\int_0^{\frac{\pi}{3}} 2 \sec t \, (3t \cos t + 3 \sin t) \, dt$

   $= \displaystyle\int_0^{\frac{\pi}{3}} (\tan t + t) \, dt$

   (c) $6 \ln 2 + \dfrac{\pi^2}{3}$

C - Sequences and series

1. $2 - \dfrac{9}{4}x - \dfrac{81}{64}x^2 - \dfrac{729}{512}x^3$

2. (a) $1 + x + \dfrac{3}{2}x^2 + \dfrac{5}{2}x^3$

   (b) $0.1 + 0.1x + 0.15x^2 + 0.25x^3$

3. (a) $a = 6, n = 2.5$

   (b) 67.5

4. (a) $\frac{1}{2}x^2 + \frac{1}{4}x^3$

   (b) $f''(x) > 0$

D - Differentiation

1. $5y + 4x + 6 = 0$

2. (a) $\dfrac{-2\sin t \cos t}{\operatorname{cosec}^2 t}$

   (b) $x + 2y = 4$

3. (a) £1980 (3 s.f.)

   (b) -801

   (c) Rate of decrease in value on 1st January 2005

4. $(y - \frac{1}{2}a) = -\dfrac{\sqrt{3}}{6}(x - 0)$

E - Integration

1. 16

2. (a) $A = -1$, $B = 3$, $C = 3$

(b) $-2 + 3\ln5$

3. $\frac{2}{9}e^3 + \frac{10}{9}$

4. (a) $\frac{1}{2}x\sin 2x + \frac{1}{4}\cos 2x + c$

(b) $\frac{x^2}{4} + \frac{x}{4}\sin 2x + \frac{1}{8}\cos 2x + k$

5. $\pi\left(\frac{13}{4}e^6 - \frac{1}{4}e^2\right)$

6. (a) $y = 0 \Rightarrow \cos t = \frac{1}{2}$, $t = \frac{\pi}{3}$ or $\frac{5\pi}{3}$

(b) Area $= \int y\,dx = \int_{\pi/3}^{5\pi/3}(1 - 2\cos t)(1 - 2\cos t)\,dt$

$= \int_{\frac{\pi}{3}}^{\frac{5\pi}{3}}(1 - 2\cos t)^2\,dt$

(c) $4\pi + 3\sqrt{3}$

7. (a) $\frac{dV}{dt} = 20 - kV$

(b) $A = \frac{20}{k}$, $B = -\frac{20}{k}$

(c) $\frac{75}{\ln2}$ cm^3

8. (a) $4\pi r^2$

(b) $\frac{1000}{4\pi r^2(2t+1)^2}$

(c) $500\left(1 - \frac{1}{2t+1}\right)$

(d) (i) 4.77 cm (ii) 0.0289 cms^{-1}

9. (a) 1.01959, 1.08239, 1.41421

(b) 0.8859

(c) 0.51 %

F - Vectors

1. (a) $a = 18$, $b = 9$

(b) (6, 10, 16)

(c) $14\sqrt{2}$

2. (a) $p = 3$, $q = 5$

(b) $\frac{1}{9}(6\mathbf{i} + 6\mathbf{j} + 3\mathbf{k})$ *o.e.*

(c) $\frac{8}{9}$

(d) (5, 3, 5)

3. (a) $\mathbf{r} = 5\mathbf{j} + 5\mathbf{k} + t(3\mathbf{i} - 3\mathbf{j} - 6\mathbf{k})$ *o.e.*

(b) $2.5\mathbf{i} + 2.5\mathbf{j}$

(c) $-3\mathbf{i} + 3\mathbf{j} + 6\mathbf{k}$

(d) $15\sqrt{3}$

Core Mathematics C4 - Practice paper

Answers to these practice papers have not been included in this book. Your teacher should have access to the mark schemes through Edexcel Online. Alternatively, visit www.examzone.co.uk to download model answers.

Paper Reference(s)

6666/01

Edexcel GCE

Core Mathematics C4
Advanced Level

Thursday 15 June 2006 – Afternoon

Time: 1 hour 30 minutes

Materials required for examination	Items included with question papers
Mathematical Formulae (Green)	Nil

Candidates may use any calculator EXCEPT those with the facility for symbolic algebra, differentiation and/or integration. Thus candidates may NOT use calculators such as the Texas Instruments TI 89, TI 92, Casio CFX 9970G, Hewlett Packard HP 48G.

Instructions to Candidates

In the boxes above, write your centre number, candidate number, your surname, initial(s) and signature.
Check that you have the correct question paper.
When a calculator is used, the answer should be given to an appropriate degree of accuracy.
You must write your answer for each question in the space following the question.

Information for Candidates

Full marks may be obtained for answers to ALL questions.
The marks for individual questions and the parts of questions are shown in round brackets: e.g. **(2)**.
There are 7 questions in this question paper.
The total mark for this paper is 75.

Advice to Candidates

You must ensure that your answers to parts of questions are clearly labelled.
You must show sufficient working to make your methods clear to the examiner. Answers without working may gain no credit.

1. A curve C is described by the equation

$$3x^2 - 2y^2 + 2x - 3y + 5 = 0.$$

Find an equation of the normal to C at the point $(0, 1)$, giving your answer in the form $ax + by + c = 0$, where a, b and c are integers.

(7)

Question 1 continued

2.
$$f(x) = \frac{3x-1}{(1-2x)^2}, \qquad |x| < \tfrac{1}{2}.$$

Given that, for $x \neq \tfrac{1}{2}$, $\dfrac{3x-1}{(1-2x)^2} = \dfrac{A}{(1-2x)} + \dfrac{B}{(1-2x)^2}$, where A and B are constants,

(a) find the values of A and B.

(3)

(b) Hence, or otherwise, find the series expansion of f(x), in ascending powers of x, up to and including the term in x^3, simplifying each term.

(6)

Question 2 continued

Question 2 continued

Leave
blank

3.

Figure 1

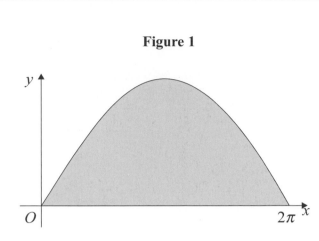

The curve with equation $y = 3\sin\dfrac{x}{2}$, $0 \leqslant x \leqslant 2\pi$, is shown in Figure 1. The finite region enclosed by the curve and the x-axis is shaded.

(a) Find, by integration, the area of the shaded region.

(3)

This region is rotated through 2π radians about the x-axis.

(b) Find the volume of the solid generated.

(6)

Question 3 continued

Question 3 continued

4.

Figure 2

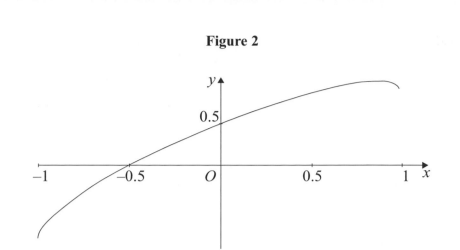

The curve shown in Figure 2 has parametric equations

$$x = \sin t, \quad y = \sin\left(t + \frac{\pi}{6}\right), \qquad -\frac{\pi}{2} < t < \frac{\pi}{2}.$$

(a) Find an equation of the tangent to the curve at the point where $t = \frac{\pi}{6}$.

(6)

(b) Show that a cartesian equation of the curve is

$$y = \frac{\sqrt{3}}{2}x + \frac{1}{2}\sqrt{(1-x^2)}, \qquad -1 < x < 1.$$

(3)

Question 4 continued

Question 4 continued

5. The point A, with coordinates $(0, a, b)$ lies on the line l_1, which has equation

$$\mathbf{r} = 6\mathbf{i} + 19\mathbf{j} - \mathbf{k} + \lambda(\mathbf{i} + 4\mathbf{j} - 2\mathbf{k}).$$

(a) Find the values of a and b.

(3)

The point P lies on l_1 and is such that OP is perpendicular to l_1, where O is the origin.

(b) Find the position vector of point P.

(6)

Given that B has coordinates $(5, 15, 1)$,

(c) show that the points A, P and B are collinear and find the ratio $AP:PB$.

(4)

Question 5 continued

6.   **Figure 3**

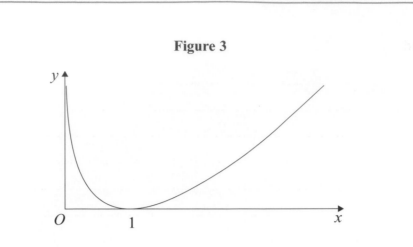

Figure 3 shows a sketch of the curve with equation $y = (x-1)\ln x, \quad x > 0$.

(a) Complete the table with the values of y corresponding to $x = 1.5$ and $x = 2.5$.

x	1	1.5	2	2.5	3
y	0		$\ln 2$		$2\ln 3$

(1)

Given that $I = \displaystyle\int_{1}^{3} (x-1)\ln x \, dx$,

(b) use the trapezium rule

    (i) with values of y at $x = 1$, 2 and 3 to find an approximate value for I to 4 significant figures,

    (ii) with values of y at $x = 1$, 1.5, 2, 2.5 and 3 to find another approximate value for I to 4 significant figures.

(5)

(c) Explain, with reference to Figure 3, why an increase in the number of values improves the accuracy of the approximation.

(1)

(d) Show, by integration, that the exact value of $\displaystyle\int_{1}^{3} (x-1)\ln x \, dx$ is $\frac{3}{2}\ln 3$.

(6)

Leave blank

Question 6 continued

Question 6 continued

Question 6 continued

Question 6 continued

7.

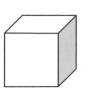

At time t seconds the length of the side of a cube is x cm, the surface area of the cube is S cm^2, and the volume of the cube is V cm^3.

The surface area of the cube is increasing at a constant rate of 8 cm^2 s^{-1}.

Show that

(a)　$\dfrac{\mathrm{d}x}{\mathrm{d}t} = \dfrac{k}{x}$, where k is a constant to be found,

(4)

(b)　$\dfrac{\mathrm{d}V}{\mathrm{d}t} = 2V^{\frac{1}{3}}$.

(4)

Given that $V = 8$ when $t = 0$,

(c)　solve the differential equation in part (b), and find the value of t when $V = 16\sqrt{2}$.

(7)

Question 7 continued

Question 7 continued

TOTAL FOR PAPER: 75 MARKS

END

Licence Agreement: *Edexcel Topic Tutor A-Level Maths, Core 3 & Core 4 Student CD-ROM* (ISBN: 978-1-84690-013-6)

Warning:

This is a legally binding agreement between You (the user) and Edexcel Limited, of One90 High Holborn, London, WC1V 7BH, United Kingdom ("Edexcel").

By retaining this Licence, any software media or accompanying written materials or carrying out any of the permitted activities You are agreeing to be bound by the terms and conditions of this Licence. If You do not agree to the terms and conditions of this Licence, do not continue to use the *Edexcel Topic Tutor Student CD-ROM* and promptly return the entire publication (this Licence and all software, written materials, packaging and any other component received with it) with Your sales receipt to Your supplier for a full refund.

Edexcel Topic Tutor Student CD-ROM consists of copyright software and data. The copyright is owned by Edexcel. You only own the disk on which the software is supplied. If You do not continue to do only what You are allowed to do as contained in this Licence you will be in breach of the Licence and Edexcel shall have the right to terminate this Licence by written notice and take action to recover from you any damages suffered by Edexcel as a result of your breach.

Yes, You can:

1. use *Edexcel Topic Tutor Student CD-ROM* on your own personal computer as a single individual user.

No, You cannot:

1. copy *Edexcel Topic Tutor Student CD-ROM* (other than making one copy for back-up purposes);
2. alter the software included on the *Edexcel Topic Tutor Student CD-ROM*, or in any way reverse engineer, decompile or create a derivative product from the contents of the database or any software included in it (except as permitted by law);

3. include any software data from *Edexeel Topic Tutor Student CD-ROM* in any other product or software materials;
4. rent, hire, lend or sell *Edexcel Topic Tutor Student CD-ROM* to any third party;
5. copy any part of the documentation except where specifically indicated otherwise;
6. use the software in any way not specified above without the prior written consent of Edexcel.

Grant of Licence:

Edexcel grants You, provided You only do what is allowed under the Yes, You can section above, and do nothing under the No, You cannot section above, a non-exclusive, non-transferable Licence to use *Edexcel Topic Tutor Student CD-ROM*.
The above terms and conditions of this Licence become operative when using *Edexcel Topic Tutor Student CD-ROM*.

Limited Warranty:

Edexcel warrants that the disk or CD-ROM on which the software is supplied is free from defects in material and workmanship in normal use for ninety (90) days from the date You receive it. This warranty is limited to You and is not transferable.

This limited warranty is void if any damage has resulted from accident, abuse, misapplication, service or modification by someone other than Edexcel. In no event shall Edexcel be liable for any damages whatsoever arising out of installation of the software, even if advised of the possibility of such damages. Edexcel will not be liable for any loss or damage of any nature suffered by any party as a result of reliance upon or reproduction of any errors in the content of the publication.

Edexcel does not warrant that the functions of the software meet Your requirements or that the media is compatible with any computer system on which it is used or that the operation of the software will be unlimited or error free. You assume responsibility for selecting the software to achieve Your intended results and for the installation of, the use of and the results obtained from the software.

Edexcel shall not be liable for any loss or damage of any kind (except for personal injury or death caused by its negligence) arising from the use of *Edexcel Topic Tutor Student CD-ROM* or from errors, deficiencies or faults therein, whether such loss or damage is caused by negligence or otherwise.

The entire liability of Edexcel and your only remedy shall be replacement free of charge of the components that do not meet this warranty.

No information or advice (oral, written or otherwise) given by Edexcel's employees or agents shall create a warranty or in any way increase the scope of this warranty.

To the extent the law permits, Edexcel disclaims all other warranties, either express or implied, including by way of example and not limitation, warranties of quality and fitness for a particular purpose in respect of *Edexcel Topic Tutor Student CD-ROM*.

Governing Law:
This Licence will be governed and construed in accordance with English law.